# BIGFOOT VS. KRAMPUS

**Disclaimer:**

The creatures in this book are not real. They are from myths. They are fun to imagine. Read the 45th Parallel Press series Magic, Myth, and Mystery to learn more about them.

45TH PARALLEL PRESS

Published in the United States of America by Cherry Lake Publishing
Ann Arbor, Michigan
www.cherrylakepublishing.com

Reading Adviser: Marla Conn, MS, Ed., Literacy specialist, Read-Ability Inc.
Book Designer: Melinda Millward

Photo Credits: © LOMAKIN/Shutterstock.com, back cover, 16; © RikoBest/Shutterstock.com, cover, 5, 10; © Budimir Jevtic/Shutterstock.com, cover, 5; © aleks1949/Shutterstock.com, 6; © United Archives GmbH/Alamy Stock Photo, 9; © Nicola Simeoni/Shutterstock.com, 12; © FooTToo/Shutterstock.com, 15; © John Zada/Alamy Stock Photo, 19; © GraphicsRF/Shutterstock.com, 19, 20; © vectortatu/Shutterstock.com, 20; © Sergio Delle Vedove/Dreamstime.com, 21; © meunierd/Shutterstock.com, 23; © Ariantolog/Shutterstock.com, 24; © Ron Rowan Photography/Shutterstock.com, 25; © kryzhov/Shutterstock.com, 25; © Science History Images/Alamy Stock Photo, 27; © anthonysp/istockphoto.com, 29

Graphic Element Credits: © studiostoks/Shutterstock.com, back cover, multiple interior pages; © infostocker/Shutterstock.com, back cover, multiple interior pages; © mxbfilms/Shutterstock.com, front cover; © MF production/Shutterstock.com, front cover, multiple interior pages; © AldanNi/Shutterstock.com, front cover, multiple interior pages; © Andrii Symonenko/Shutterstock.com, front cover, multiple interior pages; © acidmit/Shutterstock.com, front cover, multiple interior pages; © manop/Shutterstock.com, multiple interior pages; © Lina Kalina/Shutterstock.com, multiple interior pages; © mejorana/Shutterstock.com, multiple interior pages; © NoraVector/Shutterstock.com, multiple interior pages; © Smirnov Viacheslav/Shutterstock.com, multiple interior pages; © Piotr Urakau/Shutterstock.com, multiple interior pages; © IMOGI graphics/Shutterstock.com, multiple interior pages; © jirawat phueksriphan/Shutterstock.com, multiple interior pages



**45th Parallel Press** is an imprint of Cherry Lake Publishing.

Library of Congress Cataloging-in-Publication Data

Names: Loh-Hagan, Virginia, author.
Title: Bigfoot vs. Krampus / by Virginia Loh-Hagan.
Other titles: Bigfoot versus Krampus
Description: Ann Arbor, Michigan : Cherry Lake Publishing, 2020. | Series: Battle royale : lethal warriors | Includes index.
Identifiers: LCCN 2019032919 (print) | LCCN 2019032920 (ebook) | ISBN 9781534159341 (hardcover) | ISBN 9781534161641 (paperback) | ISBN 9781534160491 (pdf) | ISBN 9781534162792 (ebook)
Subjects: LCSH: Sasquatch–Juvenile literature. | Krampus (Legendary character)–Juvenile literature.
Classification: LCC QL89.2.S2 L643 2020 (print) | LCC QL89.2.S2 (ebook) | DDC 001.944–dc23
LC record available at https://lccn.loc.gov/2019032919
LC ebook record available at https://lccn.loc.gov/2019032920

Printed in the United States of America
Corporate Graphics

## About the Author

Dr. Virginia Loh-Hagan is an author, university professor, and former classroom teacher. She was interviewed on a podcast and talked about Krampus. She lives in San Diego with her very tall husband and very naughty dogs. To learn more about her, visit www.virginialoh.com.

# Table of Contents

# Introduction

Imagine a battle between Bigfoot and Krampus. Who would win? Who would lose?

Enter the world of *Battle Royale: Lethal* **Warriors**! Warriors are fighters. This is a fight to the death! The last team standing is the **victor**! Victors are winners. They get to live.

**Opponents** are fighters who compete against each other. They challenge each other. They fight with everything they've got. They use weapons. They use their special skills. They use their powers.

They're not fighting for prizes. They're not fighting for honor. They're fighting for their lives. Victory is their only option.

Let the games begin!

VS.

# BIGFOOT

Bigfoot isn't just a name for one creature. It can be used to describe a group of these creatures.

Bigfoot is a giant hairy ape-man. He has big feet. (That's how he got his name.) His feet are 24 inches (61 centimeters) long. They're 8 inches (20 cm) wide. They have thick **soles**. Soles are the bottoms of feet. His feet are special. They're built to travel rough ground. Bigfoot can travel over rocks and other sharp things. He can survive in tough weather. He can live through storms.

He has a special **gait**. Gaits are walking styles. Bigfoot walks on two legs. He walks **upright**. Upright means standing. Bigfoot takes long steps. He swings his arms widely. He can cover a lot of ground.

Bigfoot's body looks like a human's body. But he's a lot bigger. He's 7 to 10 feet (2 to 3 meters) tall. He weighs about 1,000 pounds (454 kilograms). He has many muscles. He has wide shoulders. He has long arms. He has thick legs. He also looks like an ape. He's covered in dark, shaggy hair. He has a gorilla face. He has a large forehead.

He's really strong. He has super strength. He uses his big size. He can lift heavy rocks. He can lift cars. He twists tree trunks. He hits with his fists. He twists the necks of his prey.

Bigfoot may live in small family groups.

Bigfoot is a good hunter. He has good senses. He sees well. He hears well. He smells well. He's a good **stalker**. Stalk means to follow. Bigfoot tracks his prey. He's also athletic. He runs fast. He jumps high and far. He's a good swimmer.

Bigfoot is active at night. He hides during the day. He doesn't want to be seen. He usually runs away from humans. He rarely attacks. But he does try to scare others away. He'll scream. He'll throw rocks. He'll shake trees. He'll snap branches. He'll thump his chest. He can communicate. He makes loud calls over long distances. He clicks. He grunts. He growls.

# FUN FACTS ABOUT BIGFOOT

- Jane Goodall is a famous scientist. She's an expert on chimpanzees. She's interested in Bigfoot. She said, "Maybe they don't exist, but I want them to." Studying Bigfoot is good. Experts study folktales. They study people's stories. In doing so, they discover other things.
- The "Sierra Sounds" are Bigfoot sounds. The sounds were recorded in California. They were recorded in the Sierra Nevada mountains. The sounds include grunts, snarls, and howls.
- Sasquatch are similar to Bigfoot. But they live in Canada. Sasquatch comes from the word *sasq'ets*. It means "wild men." It's from the Halkomelem language. This language is spoken by Native Americans. These Native Americans lived between Canada and Washington. Native American tribes have over 60 different words for Bigfoot.
- Chewbacca is a popular character in *Star Wars*. He was played by Peter Mayhew. He filmed in Northern California. He was in "Bigfoot country." Mayhew needed security. He didn't want to be mistaken for Bigfoot. He didn't want to get shot. His security wore bright vests.

# KRAMPUS

Krampus is popular in areas around Germany, Austria, and Norway.

Krampus is a Christmas monster. He's half-goat. He's half-**demon**. Demons are monsters. Krampus has goat horns. The horns are sharp. They curl up. Krampus is ugly. He has a long face. He has a beard. He's hairy. He has black fur. He has a long, pointed tongue. He has **fangs**. Fangs are sharp teeth. He has red, bloody eyes. He has a long tail. He has sharp claws. He has **cloven hooves**. Cloven means split. Hooves are hard covering on the feet of animals.

Some people think Krampus is Hel's son. Hel is a **Norse** goddess. Norse refers to the area around Norway. Hel is the goddess of the dead.

Krampus comes out on December 5. This is called Krampus Night. Krampus walks the streets. He visits homes. He punishes bad children. Krampus warns people he's coming. His cloven hooves make loud clicks. He rings his cowbells. He shakes his chains. His chains mean he's bound to the devil.

At Christmastime, Krampus travels with Santa. Some say he's Santa's enemy. Some say he's Santa's partner. He deals with **naughty** children. Naughty means bad. Santa deals with nice children. Krampus is **immortal**. Immortal means living forever. Krampus can be killed. But he comes back to life each December.

There are "Krampus runs" and "Krampus balls." Balls are dances.

Krampus is strong. He scares kids into behaving. He feeds off children's fears. He knows when children are bad. Good children are safe. Krampus won't harm good children.

Krampus carries bundles of sticks. Sometimes, he carries a whip. He carries a basket. The basket is strapped to his back. Krampus has sharp, scary claws. He grabs bad children. He puts children in his basket. He carries them away.

# FUN FACTS ABOUT KRAMPUS

- In the 1930s, Austria was under Fascist rule. Fascism means one leader controls everything. Believing in Krampus was banned. This means it was against the law. People couldn't celebrate Krampus Night. People who dressed as Krampus were taken to jail. In 1934, a newspaper wrote, "Krampus Disliked in Fascist Austria."

- There's a Krampus museum in Stallhofen, Austria. It has a short film about Krampus. It has about 20 Krampus masks. It has Krampus costumes and other things. It also has a Krampus cave. The story says that Krampus stays in this cave until Santa gets him out in December.

- Europeans give Krampus greeting cards. They've been doing this since the 1800s. They're called Krampuskarten. The cards read, "Gruss vom Krampus!" This means "Greetings from Krampus!" The cards have funny rhymes. They have funny poems. Krampus is on the cards. He's looming over children. He looks mean.

# CHOOSE YOUR BATTLEGROUND

Bigfoot and Krampus are fierce fighters. They're well-matched. They're both big. They both have similar powers. But they have different ways of fighting. They also have different weaknesses. So, choose your battleground carefully!

## Battleground #1: Sea

- Bigfoot comes from the Pacific Northwest area. He can swim. He's fine fighting by the shore or coast. But he prefers to be in woods. It's easier for him to hide if there are trees.

- Krampus goes all over. He travels to homes in all different places. He's from the Norway area. So, he's not afraid of the sea. But he prefers land.

Jerry Crew found a 16-inch (41 cm) footprint in 1958. The newspapers called the creature "Bigfoot."

## Battleground #2: Land

- Bigfoot lives in the wild. He loves nature. He lives off the land. He hunts and gathers. He builds homes from sticks. He sleeps in trees. He's a bit of a **nomad**. Nomad means a person who moves around a lot. But he stays away from people.

- Krampus goes to houses in cities. He goes to houses on farms. He goes all over. He likes to be around people.

## Battleground #3: Mountains

- Bigfoot lives in forests. He lives in mountains.
- Krampus lives in caves. He lives in the Alps. The Alps are mountains in Europe.

# ARMED AND DANGEROUS: WEAPONS

**Bigfoot:** Rocks are great weapons. They can be thrown. They can be rammed. Rocks can be thrown up to 82 feet (25 m) away. One hit with a rock won't kill. But many hits could do some damage. Bigfoot has to find rocks that are a good size. If the rock is too light, it won't cause enough harm. If it's too heavy, it might not make the distance.

**Krampus:** His best weapon is the "ruten." These are bundles of sticks. The bundles are made by tying up a bunch of thin sticks. The string is wrapped around the bundle several times. The technique of tying is called whipping. It looks good and is tight. Krampus mostly uses birch sticks. Birch is a type of tree. Birch is **flexible** and bumpy. Flexible means it can bend. The sticks need to be freshly cut. Birch ruten is used to hit bare legs. It stings. It makes small cuts.

# FIGHT ON!

The battle begins! It's December. It's right before Christmas. It's almost nighttime. Bigfoot and Krampus are outside a town. They're in the woods. Trees are all around. There's a campfire. But there aren't any other lights.

## Move 1:

Krampus arrives. He stomps around. His cowbells and chains make noise. They make clanking sounds. Krampus knows he's being noisy. He wants Bigfoot to know he's ready.

## Move 2:

Bigfoot hears Krampus. He hides. He stalks Krampus. He watches his every move. He slowly follows him around. But he stays far enough away not to be seen.

When scared, Bigfoot can make himself smell worse.

## Move 3:

Krampus looks for Bigfoot. He looks in all directions. But he can't see him. He just sees trees. He sniffs the air. He gets a whiff of Bigfoot. Bigfoot smells really bad.

## Move 4:

Bigfoot hides behind a tree. He throws rocks at Krampus. He throws sticks with sharp points. He runs to a different tree. He throws more rocks and sticks. He has a pile of rocks and sticks in different places.

## Move 5:

Krampus uses his bag as a shield. He covers himself. He pushes off the rocks and sticks. He moves toward Bigfoot.

There have been about 6,000 Bigfoot sightings in North America.

# LIFE SOURCE: FOOD FOR BATTLE

**Bigfoot:** Since Bigfoot is big, he needs a lot of food. He eats both animals and plants. He eats whatever he can find from the land. He hunts animals like deer. He fishes for fish. He forages for berries and fruits. Forage means to search. Bigfoot gets his water from rivers and streams. He'll even eat leftover food from humans. He'll clean up empty campsites.

**Krampus:** Krampus hibernates. Hibernates means to spend some time being inactive. Krampus only comes out around Christmas. He sleeps the rest of the year. This means he has to store up a lot of food and energy. He eats a lot of bad children in December. He eats enough to last the whole year.

## Move 6:

Bigfoot roars. He bangs on his chest. He shuffles his legs. He kicks up dirt. He lifts a tree from the ground. He throws it in front of Krampus. He wants to scare Krampus away. He wants to show how strong he is.

## Move 7:

Krampus isn't scared away. He likes the challenge. He goes after Bigfoot. He jumps over the fallen tree. He takes out his bundle of sticks. He starts whacking. He whacks left. He whacks right. He chases after Bigfoot.

Both Bigfoot and Krampus are huge male monsters.

# AND THE VICTOR IS . . .

*What are their next moves?*
*Who do you think would win?*

## Bigfoot could win if:

- He is good all year long. He can't lie or steal. Krampus only punishes bad people.
- He hides from Krampus. He needs to keep out of sight. He stays away during Christmastime. He already does a good job of not being seen.

## Krampus could win if:

- He gets dogs. Bigfoot hates dogs. Dogs are good hunters. They chase Bigfoot. They track. They attack.
- He uses a bright flashlight. Bigfoot doesn't like bright lights. Bright lights would scare Bigfoot away.

People have reported Bigfoot sightings. There are photos. There are recordings.

# Bigfoot: Top Champion

The Yeti is the alpine version of Bigfoot. Alpine refers to high mountains. The Yeti lives in the mountains in Asia. He's called the Abominable Snowman. Abominable means horrible. The Yeti has loud, high-pitched screams. He is big. But he's smaller than Bigfoot. The Yeti is strong with a lot of muscles. He weighs between 200 and 400 pounds (91 and 181 kg). He's about 6 feet (2 m) tall. He has a square head. He has white, gray, and brown fur. His fur is long and straight. The Yeti leaves big footprints in snow. Some people are scared of the Yeti. They avoid going to the mountains. They stay in town instead. But no one has found him. Sir Edmund Hillary was the first man to climb Mount Everest. In 1960, he looked for theYeti. But he didn't find him.

## Krampus: Top Champion

Frau Perchta is a female version of Krampus. Frau means married woman. Perchta is called the Christmas Witch. She's part of German folk stories. She has a beaked nose made of iron. One of her feet is very big. This foot is called a goose foot or swan foot. Some people think this means she could change into animal forms. Perchta dresses in rags. She has a cane. She carries a long knife under her skirt. She enters people's homes during the 12 days of Christmas. She travels with an army of ghosts and demons. She knows if children have been naughty or nice. She gives silver coins to good children. She puts these coins in their shoes. She hurts bad children. She hates liars and lazy people.

# Consider This!

## THINK ABOUT IT!

- How are Bigfoot and Krampus alike? How are they different? Are they more alike or different? Why do you think so?
- What skills do you have to fight Bigfoot? What skills do you have to fight Krampus?
- How are Bigfoot and Krampus good? How are they evil? Can they be both good and evil?
- How do Bigfoot and Krampus interact with humans?
- Read the 45th Parallel Press books about Bigfoot and Krampus. What more did you learn about these monsters?

## LEARN MORE!


- Brockenbrough, Martha. *Finding Bigfoot: Everything You Need to Know*. New York, NY: Feiwel and Friends, 2013.
- Dougherty, Michael. *Krampus: Shadow of Saint Nicholas*. Burbank, CA: Legendary Comics, 2015.
- Loh-Hagan, Virginia. *Bigfoot*. Ann Arbor, MI: Cherry Lake Publishing, 2016.
- Loh-Hagan, Virginia. *Krampus*. Ann Arbor, MI: Cherry Lake Publishing, 2018.


## GLOSSARY

**cloven** (KLOH-vuhn) split
**demon** (DEE-muhn) a monster
**fangs** (FANGZ) sharp, pointy teeth
**flexible** (FLEK-suh-buhl) having the ability to bend and move around easily
**gait** (GAYT) a walking style
**hooves** (HOOVZ) hard coverings of animal feet
**immortal** (ih-MOR-tuhl) being able to live forever
**naughty** (NAW-tee) not good, bad
**nomad** (NOH-mad) a person who moves around a lot
**Norse** (NORS) coming from the Norway area
**opponents** (uh-POH-nuhnts) fighters who compete against each other
**soles** (SOHLZ) the bottoms of feet
**stalker** (STAWK-ur) someone who follows or tracks someone
**upright** (UHP-rite) standing or walking on two legs
**victor** (VIK-tur) the winner
**warriors** (WOR-ee-urz) fighters

## INDEX